THE LITTLE BOOK OF THE
OCTOPUS

ASHLEY HART

summersdale

THE LITTLE BOOK OF THE OCTOPUS

Text by Kitiara Pascoe

An Hachette UK Company
www.hachette.co.uk

Summersdale Publishers
Part of Octopus Publishing Group Limited
Carmelite House
50 Victoria Embankment
LONDON
EC4Y 0DZ
UK

www.summersdale.com

The authorized representative in the EEA is Hachette Ireland, 8 Castlecourt Centre, Dublin 15, D15 XTP3, Ireland (email: info@hbgi.ie)

Printed and bound in China

ISBN: 978-1-83799-635-3

This FSC® label means
that materials used for
the product have been
responsibly sourced

MIX
Paper | Supporting
responsible forestry
FSC® C016973
FSC
www.fsc.org

Substantial discounts on bulk quantities of Summersdale books are available to corporations, professional associations and other organizations. For details contact general enquiries: telephone: +44 (0) 1243 771107 or email: enquiries@summersdale.com.

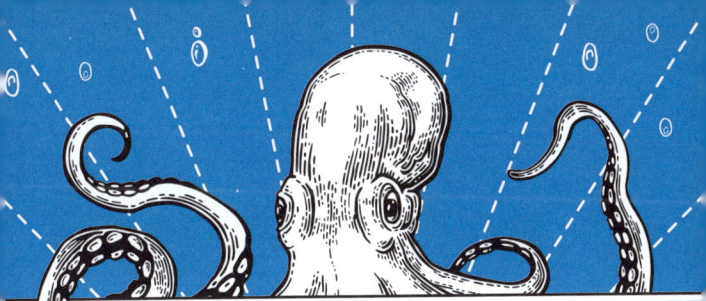

CONTENTS

🐚 INTRODUCTION 🐚

The octopus has long enamoured, mystified and scared humans as one of the most recognizable but far from well-understood animals on the planet. From tales of vast octopus-like monsters sinking ships in oceanic folklore to the heartwarming videos of these charming creatures solving complex puzzles, it's hardly surprising that they ignite such curiosity. As one of the most intelligent animals in the world, these eight-limbed molluscs can use tools, behave strategically, solve problems and learn quickly. One octopus was even tasked with predicting World Cup football results.

With around 300 species of octopus, there's significant variation in their characteristics, with just a couple of types being most commonly found in popular culture. Their habitats differ just like their species, and they live in coral reefs, deep-sea environments and even shallow water. Octopuses are shapeshifters and camouflage experts, altering their bodies in shape and colour to match rocks, seaweed and coral for safety and hunting. Adaptable is their middle name.

In this book, we'll dive deep into what makes octopuses unique, where and how they live and look at just how clever they can be. We'll also find out where octopuses have appeared in folklore and discover some of the most bizarre stories about them, as well as how they came to fill popular culture with their image.

Are you ready to learn about this beautiful, enigmatic and super-smart animal? Let's jump in and take a look under the surface.

CHAPTER ONE:
A CREATURE LIKE NO OTHER

An octopus has eight limbs – most people know that – but its number of limbs, although mighty impressive, is just the headline of this fascinating animal's biology. Seemingly more liquid than solid and almost as suited to the ocean as water is itself, it's in the octopus's physiology that the true wonder lies. This chapter explores the extraordinary facts of an octopus's biology, the unique attributes it has to survive in the ocean environment and the mind-bending intelligence that's been discovered in several of its species.

ANCIENT EVOLUTION

The octopus is part of the cephalopod class of the mollusc family, a class it shares with its cousins the squid, cuttlefish and the lesser-known nautilus. And what a class they are! Cephalopods can trace their history back to the Cambrian Period – that's 530 million years ago – and 150 million years ago evolution gifted us the octopus we know today. To put that in perspective, the extinction of the dinosaurs happened just 66 million years ago and modern humans have only been around a mere three million years. As you can see, octopuses have been around a *while*.

While it might be difficult to imagine that octopuses are actually molluscs (which typically have a hard outer shell), it becomes less surprising when we learn that their ancestors did indeed have shells. Why did evolution do away with this protective layer? Possibly to make it easier for these creatures to move to deeper environments or to enable them to escape from predators. In some cephalopods, shells are now found internally, but the octopus discarded it altogether.

DON'T MESS WITH DNA

For most animals, evolution is a long game. Random mutations occur in an animal's genes that may help or hinder its life. Over time, the useful genes are reproduced through generations to create more permanent DNA changes to that species. The DNA contains genetic instructions for how to build each life form. Changes in DNA take considerable time to occur at scale, although in some organisms like bacteria and plants, change can happen rapidly. Octopuses though, like to do things... differently.

One of the most fascinating things about the octopus is that its DNA hasn't changed much at all. In the last decade, scientists have discovered that cephalopods, including the octopus, have been making changes to their RNA (ribonucleic acid), while keeping their DNA the same.

What's RNA? Like DNA, RNA carries genetic information, but it can move much more easily. Sort of like a messenger, RNA carries this information from DNA around the body, which the body's cells use to build proteins for various things (such as making hormones). While in other organisms RNA typically gets its information from DNA, in octopuses the RNA can adapt, or be edited. So, what does this mean?

RNA editing in octopuses so far has been useful for adjusting to environmental factors such as changing water temperature. This means that when an octopus is born, perhaps its DNA makes it suited for a tropical water temperature. But if you put that octopus in cooler water, instead of dying because it's not evolved for that environment, it can edit its own RNA to affect changes in its body that make it much more suited to cooler water temperatures. This is isolated evolution within a single lifespan because while DNA is hereditary – it's passed from parent to offspring – RNA is not. Changes to one octopus's RNA won't be found in that of its offspring – they must edit their own RNA to adapt to whatever environment they find themselves in. This might sound unnecessarily challenging for young octopuses, but it's this quality that makes them one of the most adaptable and resilient animals in the world.

KEEPING DNA STEADY

No one's quite sure why octopuses and other cephalopods edit their RNA rather than rely on longer-scale DNA changes – after all, they're almost the only multicellular organisms that we have discovered taking this route – but it's likely down to simple practicality. Octopuses don't have a particularly long lifespan, so they need to learn and adapt quickly to quite hazardous environments. They're also prey for many creatures, so must be adept at defence and smart at evading different predators.

One of the fascinating side effects to editing RNA instead of DNA is that it requires the DNA to remain the same over generations. If the DNA itself changed, then it wouldn't create the same coded RNA that could affect the protein in the right way to make RNA changes. Imagine a video-game developer who creates a game (RNA) to run on a version of Microsoft's operating system (DNA), but then the operating system has a big update and the game no longer has the base it needs to run. Then the game simply isn't able to communicate with the computer's operating system.

Because of RNA editing in octopuses, their underlying DNA likely hasn't changed in a staggering amount of time, tens of millions of years. Each octopus is born with more or less the same underlying DNA as an octopus 1,000, 1 million or 10 million years ago and it adapts as necessary.

To give an example of how unusual this is, only around 1 per cent of human RNA is thought to be editable during a person's lifetime, whereas around 60 per cent of a cephalopod's RNA is thought to be editable.

THE OCTOPUS'S NEAREST AND DEAREST

Cephalopods are such a fascinating class of animals that it's worth taking a moment to check out the rest. These are the cousins of the octopus:

SQUID

There are many species of squid and in general they have eight arms and two tentacles (in cephalopods, arms have suction cups along their whole length while tentacles only have them at the end). While soft like the octopus, they have an internal non-bony skeleton. Some species can eject ink as a defensive move, and most can swim very fast. The smallest species is around 1–2 cm (½–⅘ in.), while the largest is the giant squid, growing to 10–13 m (32–43 ft) in length. They're smart, can communicate and have even been known to hunt together.

CUTTLEFISH

Highly intelligent like some species of octopus and squid, cuttlefish are widespread throughout much of the world. They can squirt ink and change the colour and texture of their skin; the last two being common ways cephalopods communicate. Unlike the others, though, cuttlefish have a unique internal hard bone. The cuttlefish's prey actually includes octopus and squid, while their predators include dolphins and seals. Their size range isn't quite as wide as the squid's but is not insignificant at 4–50 cm (1½–20 in.) with the largest being the aptly named giant cuttlefish.

NAUTILUS

Nautilus are a far more primitive cephalopod and lack the complex intelligence of their cousins, but studies have shown that they have a functional memory and can learn. They've also retained their hard outer – the only cephalopod to have done so – and have a bony body that lives inside a Fibonacci-patterned shell. Nautilus rarely reach more than 25 cm (10 in.) and their lack of evolution over millions of years has earned them the moniker of "living fossils".

THE MANY SPECIES OF OCTOPUS

We probably only see a few species of octopus in popular culture but there are actually around 300. These can range from the common giant Pacific and Caribbean reef octopus to the far more unusual mimic, dumbo and blanket octopus.

You can find different species of octopus in every ocean and there are plenty of things that tie most types together. Almost all have eight arms (there is a seven-armed octopus, literally called the seven-arm octopus, but it's more of a technicality) and most can change colour and texture. The majority of species feed on small molluscs and other small animals like crabs, and they are all soft-bodied with no bones or shell.

Octopuses can live at different depths depending on their species although almost all live on the sea floor. Some are found in the shallows, occasionally visible from the shore, while others live much deeper where even light is hard to come by.

They might sound similar, but over the next few pages we'll take a look at what makes the many species different from one another.

TWO ICONIC OCTOPUSES

Let's look at the two most recognizable octopus species to see what magic they hold.

THE COMMON OCTOPUS

Aptly named, this creature is found throughout the world. They grow from 0.3–3 m (1–3 ft) from the top of their head (the mantle) to the end of their arms – if you were going to draw an octopus, this would be it. They might be common, but they're never dull. Highly intelligent, they love to play, solve puzzles and work out problems. Unlike almost every other octopus, they've been known to walk on the seafloor on two arms.

THE GIANT PACIFIC OCTOPUS

Usually hanging out at just 5 m (16 ft) below sea level, this is the biggest species of octopus and can reach widths of 9 m (30 ft) and significant weights – the largest recorded was 270 kg (595 lb), which is similar to a pig. Known for their reddish-brown colour, these are true ocean wonders and extremely intelligent, being able to think their way through problems such as how to poach crabs from crab pots.

A WILD VARIETY OF OCTOPUSES

Let's see what other surprises await us in the unique world of these eight-limbed creatures.

THE BLUE-RINGED OCTOPUS

Possibly the jazziest, this is also the most venomous species and in fact one of the most venomous *animals* in the sea. They're recognizable by their glowing blue rings, which serve to warn off predators. They are the size of a golf ball with arms just 8 cm (3½ in.) long.

THE CARIBBEAN REEF OCTOPUS

The Caribbean reef octopus is mostly iridescent but can change colour at will. It lives a nomadic existence in warm, usually shallow waters. It can grow its mantle to around 12 cm (5 in.) and arms up to 60 cm (24 in.).

THE COCONUT OCTOPUS

Just 30 cm (12 in.) from the top of their head to the end of their arms, the coconut octopus takes its name from its favourite method of staying safe and cosy – getting inside coconut shells. Highly intelligent, this species picks up ocean debris for tools and shelter, and it has another fairly unusual trait: it walks around on two arms. Only three species of octopus are known to do this.

THE DUMBO OCTOPUS

This is actually the name of 15 species of similar octopuses who live at great depths – around 4,000 m (13,000 ft) – and as a result are almost blind. They're named after the Disney elephant because their flapping fins resemble Dumbo's ears. They have no ink and cannot change colour and less is known about their behaviour due to the depths that they live in.

THE MIMIC OCTOPUS

Most octopuses use camouflage, some expertly changing colour, texture and shape. None do it quite so well as the mimic octopus, who impersonates other animals, from flounders and sea snakes to crabs. Why? To avoid being eaten. They can grow to around 60 cm (2 ft) in length.

THE FLAPJACK OCTOPUS

Living quite deep, this tiny octopus has an amazing tactic of flattening itself to look less appetizing to predators. Their arms are webbed, making them look like a miniature umbrella. They're often confused with the Dumbo octopus.

ARMED WITH ARMS

While many books and cartoons might reference an octopus's tentacles, they actually have none. Their eight limbs are called arms, and they're lined with suction cups from base to tip, unlike tentacles, which only have suction cups at the end.

One of the most jaw-dropping facts about octopus arms is that they effectively have their own brains. That's right, each arm has its own neurological bundles called ganglia and, as a result, they can move independently. These "mini-brains" are packed with neurons; in fact, around two thirds of the creature's neurons are in these smaller brains in their arms with only one third in the central brain. This means an octopus has *nine* brains.

Thanks to these smaller brains, while an octopus is doing one thing, perhaps investigating a crevice with one or two arms, another arm might be overturning a rock or trying to open a tasty-looking shell. This makes the octopus a master of multi-tasking as its arms work to move, investigate, test surfaces and even taste simultaneously. While not proven, scientists are now considering whether these arms may also be able to sense light levels.

SUCTION SECRETS

The arms work in tandem with around 2,000 suction cups in total, each cup able to move independently to achieve the arm's goal. It's these suction cups that allow the arm to manipulate objects, hold onto things and "lick" objects or prey to taste them before the octopus decides if something is edible.

BIG BRAIN, SMALL BRAIN

While the arms can move and think independently, the central, or main, brain can also control each arm. Scientists have tested this by putting food in a place an arm could only reach by first moving out of water, preventing it from using its sensitivity to chemicals to smell the food. The octopus could still *see* where the food was and managed to send its arm to the right place, even without chemical signals in the individual arm.

A BETTER COLOUR FOR BLOOD

Octopuses clearly like to be different and so why would their blood be the same as ours? Should an octopus cut itself, it's not red that would spill: it's blue.

Mammals have red blood thanks to haemoglobin, a protein that contains iron. When oxygen attaches to this protein (to be carried around the body to the cells that need it), the heme molecule inside the protein absorbs all but red light, giving it a red colour.

Octopuses have a different protein in their blood, haemocyanin, which contains copper instead of iron. It's this copper that gives their blood a blue colour when carrying oxygen. When there's little or no oxygen present (if the octopus has died), the blood lacks any colour.

Octopuses aren't alone. Squid, spiders and scorpions also have blue running through their veins. And while this seems amazing, the octopus isn't the only creature to have blood that isn't red. Some invertebrates have purple, yellow or green blood! Let's see why it's useful to have copper instead of iron in your bloodstream.

WAIT, HOW MANY HEARTS?

Octopuses possess not one, not two, but three separate hearts, and one of the reasons is likely because of their blue blood. Unlike haemoglobin, which is fast at carrying oxygen, haemocyanin is less efficient. To ensure an octopus's cells get the oxygen they need, three hearts are required, two for taking in deoxygenated blood (brachial hearts) and one for pumping out oxygenated blood (the systemic heart). The brachial hearts push deoxygenated blood past the gills to infuse it with oxygen again, before the systemic heart pumps it onwards.

So if it's the blue blood that needs three hearts, why not just have red blood and one heart? The reason is likely due to octopuses' ability to live in very cold environments, where oxygen levels are lower. As haemoglobin requires higher oxygen levels, haemocyanin is a better fit and more efficient than haemoglobin in cold temperatures, although researchers are still trying to understand the full picture. So it seems that it's octopuses' wide distribution that requires blue blood and it's the blue blood that requires three hearts!

THE OCTOPUS HALL OF FAME

Largest – The giant Pacific octopus is the largest species, and the biggest individual ever recorded spanned 9 m (30 ft) and weighed 270 kg (595 lb). They don't usually get that big though, with 3–5 m (10–16 ft) being a more normal span and 15–50 kg (33–110 lb) being a more normal weight.

Smallest – At around 1.24 cm (½ in.) and weighing less than a gram (around the same as a small paper clip), the Wolfi octopus is the smallest in the world. It hangs out in coastal areas in the Western Pacific and lives for just six months.

Deepest – Dumbo octopuses have been found in the hadal zone – the deepest trenches of the ocean, at depths greater than 6,000 m (19,700 ft). They're found in a huge range of depths but usually around 1,000 m (3,200 ft) at the shallowest. They live where no sunlight can reach.

Most venomous – Just a single milligram of venom from a blue-ringed octopus bite can kill a person, although deaths are extremely rare and they're not

aggressive to people. There's no antidote but prompt medical care usually ensures a full recovery once the venom wears off. If they bite, it's usually because they've been disturbed and are defending themselves.

Smartest – It's hard to say but the coconut, common and giant Pacific are all extremely smart and have been observed engaging in forward planning, using tools and even remembering past events. They can all solve complex puzzles, like getting food out of a closed jar.

THE REMARKABLE CAMOUFLAGE OF OCTOPUSES

Almost everything about the octopus is remarkable, but the lengths these animals go to in the name of camouflage truly is wondrous. This is an important skill in the world of a soft-bodied creature that has predators everywhere and not too much in the way of natural defences. While different species of octopus have differing camouflage abilities, they mostly fall into two categories: colour changing and texture changing.

HOW OCTOPUSES CHANGE COLOUR

Most octopuses have chromatophores in their skin, cells that can change colour by expanding or contracting to alter their hue when the cells move en masse. This can happen extremely rapidly, within milliseconds, and is quite incredible to see.

Octopuses use their colour-changing skin very deliberately, usually in one of two ways. They might be employing it for signalling purposes, where the colour change communicates something to other octopuses or animals. This could be a warning, for instance, that

the other animals shouldn't try to eat the octopus. Or they might be using adaptive colours for camouflage, changing to blend in with the surroundings or to appear like another animal or piece of the environment such as coral. This helps them hide from predators but also lie in wait for prey.

While there are plenty of other colour-changing creatures in the world, notably the chameleon, octopuses can change their colours faster than anything else on the planet.

HOW OCTOPUSES CHANGE TEXTURE

Not every predator will be fooled by a simple colour change, so many octopuses can also alter their skin's texture from rough to smooth and everything in between. This helps them camouflage even better, as they can impersonate rocks, sand or seaweed. Octopuses use the papillae in their skin – cells that can deform – to texturize themselves at will.

Changing the texture of their skin doesn't just help when hiding from predators – they also use it to look more menacing should they get spotted and to look less menacing should they want to get closer to unsuspecting prey.

LOOKING AT LIFE THROUGH AN OCTOPUS'S EYES

While humans can be colour-blind and experience different levels of visual field, we all more or less see in the same way. Not so for octopuses. Over the 300 species, their vision differs so much that there's really no one-sentence answer to how these diverse creatures view the world.

TO SEE OR NOT TO SEE

Some octopus species such as the common and the giant Pacific have excellent vision and use it for hunting and generally going about their business. Other species, such as the glass octopus, are unlikely to have very good sight at all, probably due to living at great depths without light. The blind cirrate octopus, as its name suggests, is almost totally blind, although it can sense light.

USING SKIN... TO SEE?

While octopus species have varying levels of visual sight, their skin actually has its own form of vision. The chromatophores in their skin are responsible for

the octopus's ability to change colour, but they're also the cells that sense the colour in the first place. When exposed to different light levels, the light-sensitive proteins in these cells can turn them darker or lighter. How this aids the octopus in its accurate and rapid colour-changing, scientists have yet to figure out.

NO NEED FOR EXPENSIVE SUNGLASSES

Octopuses see polarized light – this is a type of light that moves at a specific angle, let's say, horizontally or vertically. Humans can't see like this, except through polarized sunglasses, which contain a polarizing filter that blocks a specific angle of light, usually horizontal light (such as sunlight reflected off water). Octopuses see polarized light in excellent detail, allowing them to see through water without reflections or distortions.

THE COLOUR-BLIND OCTOPUS

It appears that octopuses are more or less visually colour-blind, although there are theories that suggest their eyes (as opposed to their skin) can see some colour, some of the time. Perhaps octopuses require colour vision less as they have the benefit of seeing polarized light, which travels through water better than

some colours. One theory is that because colours dull underwater as water blocks red and orange wavelengths, octopuses can get by with a smaller range of colours.

NOT YOUR AVERAGE PUPIL

The intrigue of an octopus's vision doesn't stop there – they also have fascinating, rectangular-shaped pupils. Most of the time. In the dark, their pupils become rounder to let in more light and if they come up against a predator, they can dilate their pupils to make them bigger in an attempt to scare them off – this is known as a deimatic signal.

BLIND SPOT? WHAT BLIND SPOT?

In humans and other vertebrates, the optic nerve head sits over the retina, preventing light from hitting that part of the retina. This causes us to have a blind spot. Octopuses have retinas that sit above the nerves, which means they enjoy a full, uninterrupted view in both eyes, even when using one at a time.

LISTENING WITHOUT EARS

You're not going to find an octopus with ears any time soon, but does that mean they can't hear? For decades scientists have debated on the hearing faculties of cephalopods as they don't have an air-filled chamber like fish do, where changes in pressure help with hearing. In the 2000s, it was finally discovered that octopuses could hear, specifically in the 400 Hz–1,000 Hz (1 kHz) range, and best at 600 Hz. In comparison, humans can typically hear between 20 Hz and 20,000 Hz (20 kHz) – so as you can see, octopuses aren't going to tell you to turn your music down.

One theory about why octopuses aren't very strong in the hearing department is that they tend to live on the seafloor where there are many obstacles that would block sounds outside their small range, making it futile to be able to hear them in the first place.

THE BRAINS OF THE OCEAN

One of the first things you might learn about octopuses, apart from being eight-armed, is that they are clever. Really, really clever. In fact, octopuses are one of the most intelligent animals on the planet.

To house their smarts, they have a central brain that's so big, it beats the brain-to-body ratio of any other invertebrate and most vertebrates (excluding mammals). As an added bit of fun, their brain is doughnut-shaped with their oesophagus travelling through the middle!

We've already seen how their eight arms each have their own "mini brain", an impressive bundle of neurons that process the information and guide each arm, around 500 million in total. That's roughly the same as a dog. But while your average golden retriever is good at fetching your slippers, an octopus can do an incredible amount with its brain(s)!

Later, we'll see how some octopuses have extraordinary puzzle-solving skills, as well as how they've put their remarkable smarts to mischievous use. For now, let's take a quick look into their headline talents from the thoughtful to the physical.

AN OCTOPUS'S RÉSUMÉ

Very adaptable – These liquid-like creatures can squeeze themselves into a huge array of spaces including jars, bottles, crevices and shells.

Excellent at construction – The smartest octopus species are great at creating shelters for themselves. This means finding and carrying rocks and shells to their den, a nook under a rock, and arranging them in front so they feel secure inside.

Enjoys wide selection of games – Octopuses in aquariums and labs will often play with anything they find lying about. Play might involve pushing something through the water or shooting jets of water at it to make it move. In the wild, some octopuses have been seen playing with fish shoals, where fish group together for safety.

Has own transport – Captive octopus are masters of escape and like to go exploring. Scientists have found aquarium and lab-captive octopuses in all manner of places outside their tank, including in the neighbouring tanks with suspiciously fewer fish in them than before. Octopuses use their intelligence to work out their escape route, their shape-shifting to fit through tiny holes and their eyes to make sure nobody's watching.

🐚 FAST LEARNERS 🐚

Octopuses don't parent their offspring, so infant octopuses don't get any teaching from their parents. As they're so tasty they're always at risk of becoming prey, so it's vital that they learn everything incredibly quickly. Perhaps this is why they evolved to become so smart – they simply wouldn't survive without the ability to learn fast.

Due to the seemingly endless fascinating attributes of these creatures, they're commonly studied in labs. Scientists create mazes and puzzles for them, seeing how long it takes them to learn how to do things, such as manipulate latches and locks to reach food in a closed container. It doesn't take long!

LEARNING IN CAPTIVITY

Octopuses are known for being curious, testing things, observing their surroundings and problem solving.

There are stories of captive octopuses sneaking into other tanks, eating fish and then returning to their own. In many of these scientist-reported incidents, the octopuses have taken note of the times humans aren't

likely to be looking and chosen those moments to make their mischief.

Other stories well documented on film include octopuses opening jars (from the outside *and* the inside) and even learning to do things when they want food, such as ring a bell.

LEARNING IN THE WILD

Octopuses show incredible instances of learning in the wild too. Gloomy octopuses living near each other in Australia have been observed gathering shells and throwing them at each other, but that's nothing compared to what the blanket octopus does. Female blanket octopuses reach up to 2 m (6½ ft) long and have a long flowing cape of skin between their arms, giving them a blanket-like appearance. But the males are really tiny, less than 1.5 cm (⅝ in.) long. Small and vulnerable, male blanket octopuses will find a Portuguese Man o' War (a jellyfish-like animal), rip off a venomous tentacle that they're immune to and carry it around as a weapon.

AN OCTOPUS AND ITS INK

Leaving an inky mess behind them is a common defence tactic for almost all cephalopods. The ink is held in sacs and can be released at will, usually while the octopus is making a quick getaway. They often expel a jet of water at the same time, helping to disperse the ink to obscure the vision of the predator, buying themselves precious seconds to escape.

While obscuring vision is impressive enough, it's not the only way that octopuses can use their ink. They can also release the ink with extra mucus, meaning the ink clings together to form confusing shapes for predators. The shapes are called "pseudomorphs" (false bodies) and are created to look like the octopus itself, encouraging the predator to attack these inky beings while the octopus is busy changing colour and hiding or swimming away.

Octopuses tend to have black ink while its cousins the squid and the cuttlefish have other dark colours, and the ink is mostly made up of melanin and mucus.

REGENERATION SPECIALISTS

No octopus wants to lose an arm unnecessarily but if they do, usually through a fight with a predator, then they can grow a new one. Sounds incredible but a lost arm will re-sprout. This is especially amazing given that octopuses have a short lifespan and a further seven arms to use.

Within days of losing an arm, the octopus's body floods the injured area with cells that go about forming the beginnings of the new limb. Soon, a miniature arm grows, brought on by the arrival of stem cells, cells that are able to both replicate and transform into the type of cell that the body needs in that moment. Over the next few months, the arm continues to grow complete with its new set of suction cups until the octopus has a fully grown new arm, just as good as the one it lost.

There's nothing to suggest that an octopus can't regenerate multiple limbs over its lifetime and scientists have observed octopuses regenerating more than one at the same time, making this ability even more incredible.

INVESTIGATING THE OCTOPUS'S... BEAK?

Octopuses are so agile, in part because they're a soft-bodied animal with no bones. They do have one fairly hard body part, though, and that's the mouth. Or rather, beak.

That's right, octopuses have beaks, similar to parrots, and if you can't quite picture that it's understandable you can't visualize their beaks, because they're mostly hidden away on their undersides. Much like an elephant uses its trunk to pick up food and pass it into its mouth beneath, an octopus's beak is at the base of its arms.

Let's look at the mouth of an octopus to see why it's so useful.

THE BEAK

Having a hard beak is particularly important for octopuses as they eat other molluscs, which tend to have tough shells. They can also use their beaks to cut flesh, breaking food down into manageable sizes. When not munching something, the octopus retracts its beak inside its body so it's usually invisible.

THE RADULA

Inside an octopus's mouth is the radula, a tongue-like structure covered in tough barbs. This is capable of *drilling* into the shells of molluscs to reach the body inside. This is why some empty shells have tiny holes near the apex – it's a radula-made hole that dislodges the prey's securing muscle so the octopus can pull it out and eat it. Clever, eh?

THE VENOM

Most octopuses are actually venomous and use poison to paralyze their prey, making them easy to eat without a fight. They use the radula for this as well, injecting the neurotoxin into the flesh of prey. Octopus handlers have been bitten in the past, although it's not usually a defence tactic as the mouth is in such a vulnerable place. Most octopus venom has no effect on humans, who tend to be too big for the toxin to be effective, but some octopus venom can paralyze the affected area briefly or, in the case of the blue-ringed octopus, kill the person if they don't receive medical attention quickly.

CHAPTER TWO:
THE LIFE OF THE OCTOPUS

Living in oceans full of challenges, the octopus has plenty of tricks and skills up its eight sleeves to keep itself both occupied and alive. From staying safe from predators to being a predator itself, each octopus must use its quick learning ability and immediate environment to thrive in the underwater world. We've already seen the wonders of octopus biology and in this chapter, we'll watch it in action. We'll be seeing how octopuses live their lives from birth to death, from eating to playing, and from deep sea to coral reef.

🐚 A GLOBAL ANIMAL 🐚

No matter which ocean you look in, you'll find an octopus. These highly adaptable animals are found throughout the world's oceans and commonly along coastal waters. There are some species that live in colder waters but octopuses are particularly prolific in warmer, sub-tropical waters.

Not only are they in every ocean, they're also at every depth. Plenty of species live in shallow water, sometimes inches below the surface. Others will live further down but still in spots with abundant sunlight and colourful corals. There are also deep-sea octopuses that never come to the surface willingly, living in the deepest abysses, thousands of metres down.

As their habitats and water temperatures differ, so do the octopuses themselves and where the East Pacific red octopus will thrive, the deep-swimming blanket octopus never could. Why not look up your local octopus species, and consider how close you might be to one? If you live near a coast, chances are you have a neighbourhood octopus species.

The only waters octopuses haven't conquered are freshwater bodies, so you won't find any in your local lake.

THE BIRTH OF AN OCTOPUS

A baby octopus hatches from an egg that could've been incubating for months or even years, depending on the species. By the time the octopus hatches, its mother will have died or be close to death so there's no chance for her to pass down some wisdom. A baby octopus will be surrounded by thousands of other eggs and hatchlings, although the survival rate in the wild is thought to be about 1 per cent.

PLANKTONIC OCTOPUSES

For many species of octopus, they start life as planktonic hatchlings. This means that they're not strong enough to swim against a current and so drift about in the water as they grow bigger. They eat zooplankton (tiny animals not strong enough to swim against currents) and other miniature floating invertebrates. After weeks or months, they grow big enough to settle on the seabed as young octopuses and begin their lives in earnest. During their planktonic phase, they're extremely vulnerable and many don't survive.

BENTHIC OCTOPUSES

Some species, including the Californian blue-spot, the Caribbean reef and deeper sea species, emerge from larger eggs more fully formed. As they hatch in a bigger form, they can immediately live on the seafloor and begin their rapid growth to full size.

SOLITARY INFANTS

One of the strange things about octopuses is that not only are they solitary animals (not itself unusual in the animal kingdom) but both parents die before their eggs hatch, so even their hatchlings are solitary. When an octopus gets big enough to find its own place on the seafloor, it's completely alone and must learn everything from scratch, including how to eat enough to sustain its super-fast growth. The giant Pacific octopus doubles its bodyweight every two or three months while other species grow around 5 per cent per day when very young.

HOW TO MAKE A
HOUSE A HOME

What an octopus considers a suitable place to snooze depends on the species, so let's take a closer peek at some of the most common places you might find an octopus lounging around.

IN A DEN

Many octopuses spend a significant amount of time in a den, which is a crevice or space under a rock that they feel safe in. The nook doesn't have to be perfect when they first find it; they're happy to gather rocks and shells to build out its walls and make it more secure. Sometimes they'll even add a shell door. The chosen base might stay as an octopus's home for considerable periods of time, or they might decide to look for a new one regularly – just like some humans do.

IN A SHELL

The coconut octopus is a small species that likes sheltering in shells so much it's named after one. These cute little creatures roam around finding and even collecting shells to nip into when they need shelter. They don't only use coconut cases; they will collect any shell they come across if they think they can make use of it. When truly in the mood for coconuts, they can be found inside two halves of a shell, effectively turning themselves into a well-protected ball.

IN LITTER

It's a sad fact that the world's oceans suffer from a significant amount of litter left by humans, but octopuses are good at making the most of a bad situation. Octopuses have been found hanging out in jars, bottles and other human detritus that has made its way to the seabed.

IN OPEN WATER

Unusually for these creatures, the blanket octopus doesn't keep a den at all, and instead spends its time floating about in deep but open water. This octopus has notable defence tactics, as we'll explore later.

LIVING IN THE DEEP

While it's cute to watch videos of octopuses wandering about the seafloor under dappled sunlight collecting shells, there are plenty of species who live in very different surroundings, deep in the ocean and far from sunlight.

Some octopuses, such as the Dumbo octopus, live at such great depths that we know very little about their habitats and how they interact with them. The beautiful glass octopus is another deep-sea dweller, living at 1,000 m (3,200 ft), and is so rarely seen that every sighting adds considerable new knowledge about them. These species may not even have dens in the same way shallower octopuses do, simply because they have far fewer predators.

In early 2024, a team of scientists announced they'd found a further four species of deep-living octopuses, one of which was brooding eggs near a hydrothermal vent in otherwise extremely cold water. These octopuses were nesting at 3,000 m (10,000 ft) in the company of others of the same species, and not hidden away in crevices like many octopuses would, showing that different environments can significantly impact behaviour.

🐚 AQUARIUM LIFE 🐚

You'll find octopuses in aquariums all over the world, usually those from smart species and those that are able to cope in captivity. Because octopuses are solitary creatures, they can be easier to keep than more social species. That doesn't mean aquariums or research centres have an easy time with them. As we've seen, octopuses can and will escape anything but the most secure tank, and they need games to keep them occupied – smart animals need stimulation, otherwise they can get bored and ill.

There are some international laws about keeping octopuses as pets. Despite that, many specialists do just that. Keeping an octopus, aside from the ethical issues of keeping such an intelligent creature in captivity, is not easy. They require large tanks, high-quality crab and shrimp to eat and considerable stimulation to stay healthy. In addition, they will attempt to escape often, even causing chaos and breaking their environment to do so. In some aquariums, octopuses are kept after being rescued from fishing nets, rather than being specifically acquired for the purpose of captivity.

WHAT TO SERVE AN OCTOPUS FOR DINNER

Octopuses love to eat and can be extremely cunning in how they go about finding and catching their food. Across different species, octopuses will eat crab, shrimp, worms, lobster, clams, mussels and even other octopuses. But it's not so much *what* an octopus eats but *how* it gets its dinner which is so fascinating. As octopuses must learn everything themselves from birth, different individuals will come up with different ways to get a snack. Let's take a look at some tactics octopuses have been known to employ.

DRILLING SHELLS

Creatures like clams and mussels have a superb defence mechanism – they simply clamp their shells shut when they're under attack and that keeps them safe. Well, unless it's an octopus that's doing the attacking, that is. We've already seen in the previous chapter how octopuses use their radula to drill into shells and that's exactly how they can bypass the clamp defence. They drill a hole, inject their venom to paralyze their prey and that's enough to relax the animal so they can pry the shell open and feast.

AMBUSHING

Masters of disguise, octopuses can ambush prey by changing colour and texture to match the prey's environment. They then wait until the animal is close enough for them to pounce. One species takes this a step further: the larger Pacific striped octopus is known to stretch out a single arm, curl the end around and tap prey from the opposite side so they flee into its waiting arms. Classic prank.

STALKING

Most species hunt in the evening or at night but the day octopus, otherwise known as the big blue, will spend the daylight hours stalking prey about the seabed, using its outstanding camouflage to mimic rocks, coral and seaweed.

OPEN-WATER POUNCING

The little Dumbo octopuses spend most of their time swimming about at great depth and wandering about the seabed. As a result, their prey is usually small animals such as copepods, which are tiny crustaceans that often float through water. Dumbo octopuses vacuum them up or pounce on larger (but still tiny) prey.

SHEER INGENUITY

We've already heard how captive octopuses are escape artists and they'll use these skills for catching extra snacks, food that is most likely living in another aquarium exhibit or a research tank. Smart octopuses (often the kind aquariums and research organizations keep), will observe potential prey, make a plan and do their utmost to figure out a way to get it. For researchers, this habit is useful when creating food-based puzzles but not so useful when they arrive in the morning to find the octopus has eaten other study animals. Whether some of the more elaborate tales of octopuses sneaking into other tanks and returning to their own are scientists' urban legends or not is up for debate.

🐚 LIQUID OR SOLID? 🐚

There's a common joke that cats are liquid, but they've got nothing on octopuses. Thanks to having no bones at all, even large octopuses can fit through unimaginably small holes, feeding one arm through at a time and feeling what's on the other side. With no hard outer shell, it's easy for an octopus to squeeze itself through tiny gaps and reform into its recognizable shape on the other side.

This amazing feat is useful for staying away from predators, as octopuses can retreat into small spaces and alter their shape to fit the shelter. It's also useful for finding and reaching out-of-the-way prey – large octopuses have even been known to break into fishing boats to reach crab holds.

If you can't quite imagine it, just look online for videos of octopuses squeezing through small holes. It really must be seen to be believed just how clever they are at contorting themselves to get where they want to go.

SOLITARY WITH A SIDE OF SOCIAL

Almost all octopuses are anti-social and prefer their own company to mixing with anyone else – hey, we've all been there sometimes. They're alone from the moment they hatch and drift away from their fellow eggs and must be entirely self-sufficient throughout their lives. Most of them are also territorial, so while they may live in the neighbourhood of other octopuses, they'll usually defend their territories and stay well clear of each other. This behaviour is even seen in captivity where octopuses might share a tank – they'll isolate themselves in the furthest spot from other octopuses.

They're even anti-social when it comes to mating, as we'll see. Most species of octopus mate relatively briefly and the male makes a hasty escape lest the female decides to eat him for dinner.

However, when is anything ever straightforward when it comes to the octopus? There are increasing discoveries about the social lives of octopuses and scientists are beginning to find out that some are a little more up for a chat than previously thought.

UNUSUALLY FRIENDLY OCTOPUSES

The common Sydney octopus – or gloomy octopus – might not be so gloomy after all. It was discovered in 2017 that at least two separate groups of them were living together in a small area and interacting. Researchers theorize that it may be because there were limited shelters in the area, forcing them into close proximity, but for the most part, there's no explanation. The larger Pacific striped octopus is another species known for hanging around in groups and even sharing dens.

AN OCTOPUS'S PREDATORS

The reason why octopuses are so excellent at squeezing through phenomenally small spaces is that they don't have any bones. They're made up of mostly muscle instead so their flesh is high in protein, and it's these two facts that are partially responsible for the octopus being one of the tastiest things in the ocean by pretty much everything else's standards. If a marine animal is a meat eater, it probably wants to eat an octopus.

This is problematic for our eight-armed friends as it means they have a lot of predators, including particularly ferocious ones like sharks. Their other enemies include sperm whales (capable of eating the seven-arm octopus, one of the largest species in the world), moray eels, seals and sea otters. Birds will also dive for octopuses should they see one and fish can also be predators.

As many species of octopus start off as zooplankton, floating around as newly hatched minuscule octopuses, their predators at that stage are just about anything in the sea with a mouth. This is one of the reasons that only around 1 per cent of octopuses make it to adulthood.

INTO THE JAWS OF A SHARK

Sharks are one of the most threatening predators an octopus can encounter. Armed with an excellent sense of smell and several rows of sharp teeth, sharks will locate an octopus and do their best to pull it out from whichever nook it's holed up in. Luckily for octopuses, their super-squishiness means they can often hide deep enough under rocks to evade such capture. Without a good enough shelter though, they have little defence against sharks.

ONTO THE PLATE OF A SEAFOOD RESTAURANT

Of course, a major predator of the octopus is the human. Every year, around 370,000 tons of octopuses are caught and sold commercially across over 100 species. Vast quantities are also caught as bycatch in fishing activities linked to other types of seafood. They're typically caught using pots, which they naturally like to climb into for shelter.

SO MUCH TO DO, SO LITTLE TIME

We've already touched on the short lives of octopuses but just how short are we talking? Surely such smart creatures take a while to grow up, learn all their amazing skills and get to grow old? Well, not so much. Octopuses live for between six months and five years, depending on the species, with many living just one to two years.

The giant Pacific octopus is the largest (along with the seven-arm octopus) and lives the longest, usually between three and five years. The common octopus, another super-smart species, only lives between 12 and 18 months while the tiny Wolfi octopus only lives for around six months.

One of the reasons why octopuses live a relatively short time is that almost every species dies once it's reproduced, a phenomenon called semelparity, which means that an animal only has one reproductive cycle before it dies. Octopuses experience senescence once they've bred, which is a degeneration of their cells resulting in death. We'll look a little more deeply into this seemingly odd fact in a few pages' time.

UNLIKELY FRIENDSHIPS

We know that octopuses are largely solitary and don't hang out too much with other octopuses, except perhaps in the case of Australia's gloomy octopuses and the larger Pacific striped octopus, but what about other relationships?

Most of the time, octopuses are either trying to eat or not *be* eaten, so time for social interactions is limited. When it comes to humans, however, numerous divers and researchers have reported interacting with octopuses in a more curiosity-based way. In the case of the diver in the Netflix documentary, *My Octopus Teacher*, something barely short of a friendship seemed to have been made between a human and a common octopus. We'll look more into amazing stories about how octopuses have interacted with humans in the next chapters.

It's not just humans that octopuses will strike up relationships with: day octopuses have been seen hunting collaboratively with fish on coral reefs. Sometimes fish will show an octopus where prey is and other times an octopus might frighten prey from where it was hiding, so the fish can catch it.

THE HUMAN IMPACT ON OCTOPUSES

There are some deep-sea octopuses that have lives almost entirely unaffected by the human world, principally because they live at largely unexplored depths. But for many species of octopus, humanity does have some effects on their lives.

POLLUTION

Vast quantities of plastic, metal and glass litter ends up on the ocean floor, disrupting the habitats of octopuses. One of the main ways this affects their behaviour is that octopuses adapt to use waste, like bottles and cans, as shelters. One species of recently discovered octopus, the pygmy octopus, has only ever been seen using litter as shelter, never shells. This might be disheartening but the pygmy octopus is just being resourceful. They don't know that these objects can harm them due to plastic degradation and heavy metal toxicity.

PREY POPULATIONS

Overfishing, bottom trawling and plastic pollution can all have a negative impact on the populations of animals that octopuses feed on, meaning food is harder to come by.

The animals that octopuses do consume are also now likely to have microplastics inside them, which may not have affected such small creatures, but the octopuses end up consuming considerably larger quantities of them, which could impact their health and digestion.

BYCATCH

Becoming the victim of bycatch is a huge risk for many ocean species including octopuses. This is when a non-target animal is caught in the nets, lines or pots of a fishing operation and cannot be used in the meat industry. Octopuses are fragile and easily damaged, so many caught by accident won't survive or will come away with significant injuries, making them vulnerable to predators.

CLIMATE CHANGE

The oceans have warmed significantly since the industrial revolution, and this has a considerable impact on octopuses. While they're an adaptable bunch living all over the world, even a tiny increase in water temperature can affect their health. Warmer seas mean octopuses expend more energy just living as the rise in temperature pushes their metabolic rate up and can cause them to hatch prematurely, negatively impacting their already low survival rates. Climate change is also largely responsible for coral bleaching, where huge swathes of reefs die. Without healthy coral, fish and other animals leave the reefs, meaning octopuses have nothing to eat and nowhere healthy to live.

FISHING

While octopuses are common victims of bycatch, they're also fished deliberately and extensively. While many countries have fishing quotas for octopuses – annual limits that theoretically prevent them from being overfished – this doesn't mean that octopus populations will stay stable. Fishing doesn't happen in isolation and with so many other problems caused by humans that also affect ocean environments, some octopus species may find themselves in trouble.

It's vital that we do what we can to reduce our impact on the oceans, not just to protect octopuses but to preserve the planet for all people and animals in the future. As we've seen here, waste doesn't just disappear when it goes into the ocean; it pollutes marine habitats and disrupts sea life.

THE MATING GAME

Given that they are such anti-social animals, it isn't surprising that octopuses don't have a strong bond with their mates. What's more, a female octopus may eat the male if she gets hungry while he's close by.

Male octopuses have a hectocotylus, one arm that differs from the other seven. It is usually curled up safely underneath his body. During mating, it uncurls to become the conduit by which spermatophores (sperm packets) travel across and into the female's mantle cavity via an opening near one of her eyes. In some species, these sperm packets can be very long – a whole metre in the giant Pacific octopus species!

Once the male has deposited his spermatophores, he makes a quick exit lest he be eaten. The female doesn't necessarily have fertilized eggs at this point and instead she may store the sperm until she lays eggs. In the case of blanket octopuses, the male literally hands over his hectocotylus for the female to keep until she lays eggs.

THE MARINE FOOD CHAIN

Everything has its place in the food chain – where do octopuses belong?

Photoautotrophs – This level of the food chain is where all the microorganisms of the ocean sit. This includes phytoplankton (tiny plant-like organisms), which can photosynthesize and produce oxygen. While microscopic, they exist in such quantities in the upper levels of the sea that they produce over half the world's oxygen!

Herbivores – These guys can be big or small and range from zooplankton (tiny organisms that drift in the sea) to big animals like manatees and green turtles. They only eat plant stuffs like seaweed and sea grass.

Carnivores – This level includes all the meat eaters of the ocean, including octopuses. Their diets vary, usually depending on their size. Smaller carnivores will eat zooplankton (tiny animals that float in the sea) and other small creatures, while they themselves will be eaten by larger carnivores, like octopuses.

Apex predators – Here we have the predators that tend to be big, strong and long-lived. They don't have many, if any, predators themselves and include animals like dolphins, sharks, penguins and seals.

THE IMPORTANCE OF THE OCTOPUS

While many of us love octopuses and their charming intelligence, they are an important part of the food chain in both directions. When the food chain is balanced, there aren't too few or too many of any particular animals and the marine world is in harmony. Everything needs to eat, and most things, unfortunately, will at some point probably be eaten – even sharks are eaten by bigger sharks.

Octopuses are an important food for moray eels, sea otters and seals among other predators. Because they're full of protein and fairly plentiful, these animals love to seek them out so they can get enough to eat and feed their young.

The octopus's vital role in the food chain is another reason why we should be protecting octopus populations – they're crucial for other species to survive and thrive.

HOW AN OCTOPUS LAYS ITS EGGS

Almost all species of octopus only lay eggs once because they're a largely semelparous animal, meaning they die soon after reproducing. Researchers aren't sure why octopus species so consistently die after brooding their eggs but there are a few theories. One is to keep the population stable by preventing the mature octopuses eating their own or each other's young.

Males spend much of their short adult lives looking for a female to mate with and once they do, they immediately leave to avoid being eaten (they may escape becoming dinner, but they can only mate once and soon begin to die). The female then stores the spermatophores until she's ready to lay her eggs. This could take weeks or months but once she's laid eggs, there's no going back for her.

The female octopus is a truly dedicated mother. In the case of many species, the female will lay thousands, tens of thousands or even hundreds of thousands of eggs one at a time and do something incredible: she'll take each egg and glue it to another using mucus. She'll keep repeating this process until she has long braids of eggs. When the braid is long enough, she'll begin a new one.

The female of some species will store these braids in a den, safely away from any predators, and she'll stay with them at all times, never leaving nor feeding herself. Other species will wrap the braids around their bodies, taking them with them through the ocean. Others still, particularly in the abyssal deeps, will lay eggs onto rocks and stand guard.

Females will blow bubbles over the eggs during the brooding time, keeping them oxygenated and clean of dirt or algae. Some mothers need to nest for weeks before the eggs hatch, whereas the giant Pacific octopus broods over hers for ten months. Nothing beats the nesting time of the *Graneledone boreopacifica* though – females of this species have been observed brooding their eggs for four and a half years! This happens at great depths, where the extreme cold slows down the incubation process.

ALL GOOD THINGS COME TO AN END

Octopuses don't have particularly glamorous deaths, and it almost seems unnecessarily sad. When it comes to the end of an octopus's life, scientists have relatively little understanding of why they die in the way they do, although research is moving forward in this sphere.

After mating in the male's case, and after brooding eggs in the female's case, octopuses enter a phase in their lives called senescence. This is a stage of life where their cells stop repairing themselves and the octopus stops eating. This means that its body begins to degenerate, resulting in death.

For the male, they quickly fall into senescence after they've mated with a female – meaning they can only mate once – and die within weeks, sometimes months. During this time, they become weak and can't defend themselves as easily, so are easier for predators to catch.

For the female, senescence begins when she lays her eggs and continues throughout the brooding period, which could be months or, as we've just seen, years. She never eats during this time – the same goes for octopuses that brood for significant periods – and

instead is single-minded about taking care of her eggs as her body falls apart.

In both male and female octopuses, senescence also often includes a final period of self-destruction where octopuses have even been observed eating their own arms. This is during the time when the octopuses are very near death and can't feel pain any more nor make logical decisions.

Female octopuses will die just before or around the time their eggs hatch, letting her young start their lives completely alone as she did.

In recent years, researchers have discovered what triggers the senescence in the first place – the optic glands. These glands are responsible for causing the biochemical changes that shut down an octopus's dietary system and prevent it from eating. Why? Nobody knows but it might be to keep the population in balance. While it sounds beneficial for octopuses to live longer and breed more, if there are too many octopuses then there wouldn't be enough food, and they would die in other ways.

CHAPTER THREE:
MYTHS, LEGENDS AND MYSTERIES

In many cultures, an octopus's reputation truly precedes it. Sometimes it's the creature's masterful intelligence that people first hear about, but for many, it's more likely to be one of the incredible myths and legends that surrounds this fascinating animal.

You can find octopus-based tales the world over, with so many bearing such similarity that it's hard to believe there's not significant truth behind even the most absurd or unbelievable. Some cultures have stories about octopuses that date back a thousand years!

In this chapter, we'll dive into the tantalizing tales and mysterious mythologies surrounding octopuses. We'll also see some of the more amazing examples of how individual animals have interacted with humans.

WHAT'S SO SPECIAL ABOUT THE OCTOPUS?

We're about to look at the many traditions and fables surrounding octopuses but first, let's consider this question: why the octopus at all?

There are plenty of animals out there that are bigger, scarier and more dangerous than octopuses. There are even some that are more bizarre. And yet, you don't hear centuries-old talks of the Big Man-Eating Duck Billed Platypus, or the Monster Mouse of Sweden, or the Giant-Footed Anteater. Just to be clear, all of those are made up.

One reason for the octopus being the subject of so many tales could be its environment. We *still* don't know much about the underwater world, and we certainly didn't know much about it at all hundreds of years ago. The sea has historically felt like quite a mysterious place as well as a dangerous one: many sailors have died at sea throughout history. The ocean also has phenomena caused by weather that sailors of old did not understand. Put this together with the alien-like eight arms and you have yourself a perfect scary story character.

THE LEGEND OF
THE KRAKEN

Traceable back to the thirteenth century in Scandinavian folklore, tales of the Kraken have peppered seafarers' tales for centuries. In the Norse sagas, there were mentions of gigantic monsters appearing in the North Atlantic, large enough to take down ships. Stories of enormous tentacles that would wrap around hulls and pull them to watery graves would've terrified sailors of the time.

The Kraken is often depicted as a giant squid – a real species that can reach up to around 13 m (43 ft) long and might be understandably frightening for mariners throughout history. Over the centuries, it's also been likened to a gigantic octopus, which is the more common depiction in stories today. As we've seen, the largest octopus known to humans is the giant Pacific octopus, reaching around 9 m (30 ft) across and weighing up to 70 kg (154 lb); not enough to sink a ship, but not something that ancient seafarers would've easily dismissed.

It's thought that the stories may have come about from sailors seeing carcasses of these large creatures floating on the sea or washing up on beaches.

Tentacled (or should we say, many-armed) creatures are so unusual that it's not hard to imagine how easily a mariner's excited recollection of a mysterious beast could catch on and be carried through the centuries.

THE POWER OF AN IDEA

Hundreds of years ago sailors could've been at sea for months, and even years, in complete isolation that is hard for us to contemplate today. With wild storms, churning seas and an understandable fear of the unfamiliar, seafarers often leaned on myths and superstitions to explain the things they saw and felt. Stories spread through countries and centuries, cementing the idea of a giant, tentacled sea monster lurking in the depths. Considering that we've explored relatively little of the world beneath the ocean even today, perhaps it's still this sense of the unknown that keeps the Kraken in popular culture.

AKKOROKAMUI

Now we move on to the myths of the Ainu people, a native Japanese ethnic group originating from the north of Japan. The Ainu live in close harmony with nature and their myths are tightly knit with the land around them. Akkorokamui is the Japanese name for an octopus that was born in Ainu folklore; its Ainu name is Atkorkamuy.

It's said that there was once an enormous spider living in the mountains named Yaushikep, so big that its body covered an entire hectare. One day, the spider attacked a nearby village and the villagers begged the gods for help to escape the beast's terrible rampage. In answer, the sea god Rep-un-Kamuy captured the giant spider and submerged it in the sea at Uchiura Bay, a huge bay south-west of Sapporo.

Underwater, Yaushikep transformed into a gigantic octopus, the same size as the spider and the same deep-red colour. From that point, villagers in the area called the creatures Atkorkamuy, or Akkorokamui.

AKKOROKAMUI THE MONSTER

Akkorokamui was said to be able to swallow ships whole and turn the sky and water red with its reflection, to the point where fishermen in history would avoid the bay whenever there was a red tint in the sky. In Ainu folklore, the octopus had a strong, noxious smell whenever it appeared.

AKKOROKAMUI IN SHINTO

Shinto is a nature-based religion of Japanese origin and features a whole host of animal-esque deities and mythological creatures. Akkorokamui has found its way into Shinto as well, and is a well-known figure, with shrines to the creature dotted around the islands. Thanks to octopuses' ability to regrow limbs, Akkorokamui is often the figure to go to for healing. Practitioners will give offerings of seafood when asking for Akkorokamui's medical help. Mental purification is another reason one might turn to a shrine to this mystical octopus.

KANALOA

A mere 6,500 km (4,000 miles) south-east of Japan lies Hawaii and the sea god Kanaloa. Appearing in many forms, one of the most popular representations of this powerful god is as an octopus.

Kanaloa is one of the four primary Hawaiian gods, including Kane, his brother. They're often talked about together, as they represent two opposites; while Kane stands for order, Kanaloa represents chaos.

Kanaloa is a sea god, incredibly important in local folklore, not least because Hawaii is an archipelago where its people have always relied upon, and feared, the ocean surrounding them. He is a shapeshifter as well, taking many forms from dolphin and whale to wave and octopus. Among other things, in Hawaiian history he represents darkness and mystery, power and the unseen. Just like the deep ocean he lives in, he's multifaceted.

Depending on the myth, Kanaloa is variously depicted as Kane's twin or his son. No matter which myth you study, these two gods are always innately joined and co-creating balance in the natural world. Like the yin and yang in Taoism and the dual opposite deities in many religions, they represent the duality of the world.

When it comes to command over the sea, no one wields more power in Hawaiian myths than Kanaloa. He is the god of navigation and seafaring, keeping sailors safe as they voyage in the vast Pacific. He also controls the tides and watches over underwater life. Sailors will pay respects to him when they set off, but he's also invoked in healing rituals, thanks to his associations with well-being. Finally, Kanaloa is the god of the underworld, although in Polynesian culture, that position is given to an ancient Hawaiian god named Milu.

TE WHEKE O MUTURANGI

Still in the Pacific, we come to the tale of Te Wheke o Muturangi (Muturangi's octopus). This is a Māori story from New Zealand and it begins in the land of Hawaiki, the origin lands of Pacific Islanders in Polynesian mythology.

The story goes that on an island in Hawaiki, the bait from the fishermen's lines were going missing. This had never happened before, and they were sure they hadn't done anything to displease the gods. As their tribe relied on fishing, a fisherman named Kupe decided to get to the bottom of the problem. He noticed that the bait-less hooks had octopus slime on them, and sought out a banished priest named Muturangi, who was said to have a gigantic octopus as a pet named Te Wheke o Muturangi.

Muturangi had taught the octopus – wheke in Māori – to catch the fish in the fishermen's nets and steal bait from their lines, so that it could bring food back to him in revenge for having been banished from the village. When Kupe discovered this, he decided to find and kill the great octopus.

He set about building a huge canoe (*waka*) with the other villagers and prepared to go to sea. With warriors from the village, Kupe cast a spell on Te Wheke o Muturangi, ensuring that it could no longer swim deeply and hide. Forced to the surface, the octopus fled across the ocean and for weeks Kupe and his men rowed in pursuit. Eventually, they spotted some islands covered in clouds and they named the archipelago Aotearoa – the land of the long white cloud.

Kupe and the warriors chased the octopus around the north island, into bays and along the coastline. Eventually, they caught up with it in the Cook Strait and finally killed the octopus and left its eyeballs on rocks in the strait – *Ngāwhatu-kai-ponu* (the eyes that saw) – which are today's Brother Islands to the north-west of Wellington, off Arapaoa Island.

This is not only an important octopus story, it's the Māori tale for how Polynesians discovered New Zealand in the first place.

KOROMODAKO

Japanese folklore has a range of supernatural creatures, often taking the form of animals (but sometimes also demons and ghosts), called *yōkai*. Over the centuries, much natural phenomena has been attributed to these creatures as a way of understanding them.

One *yōkai* is the Koromodako, or cloth octopus. The Koromodako's usual appearance isn't anything special. Like many real octopuses, the males are described as being very small, just an inch or so long, while the females are around five times larger, so nothing particularly unnerving. They are said to live in the Sea of Japan between the prefectures of Kyoto and Fukui.

So if these supernatural beings just resemble normal octopuses, what makes them so special? Well, when threatened, they instantly increase in size massively, stretch their skin between their arms like a huge cloth (hence "cloth octopus") and can swallow entire ships. After the creature has consumed whatever disturbed it, it returns to its usual, unassuming size. If there was ever a story that emphasized the unexpectedness of octopuses, surely this is it.

KUMUGWE'S OCTOPUS

The indigenous people of the Pacific Northwest, namely the Kwakwaka'wakw and the Nuxalk peoples, also have an octopus in their mythologies. Responsible for the tides and the ocean world, the sea god Kumugwe lives in an underwater palace where sea lions act as columns, holding up the building. He is said to consume human eyes should he find someone washed into the sea, but he's also a benevolent figure, a healer.

Heroes that reach his palace are rewarded richly but first they have to get past his guard... an octopus.

Sometimes referred to as Komokwa, Kumugwe is a frequent figure in the area's art, particularly in the form of a mask. He's sometimes represented as an octopus and other times with octopus suction cups on his face, along with other marine life.

It's no surprise that an octopus should feature in Kwakwaka'wakw and Nuxalk stories as these groups originate (and still largely live) in the coastal and island areas of Canada's British Columbia, where local cultures live alongside the sea.

PLUTARCH'S OCTOPUSES

In ancient Greece, we discover a number of retellings of an imagined banquet for important guests and seven storytellers. This is the *Banquet of the Seven Sages*. The storytellers featured were real historical figures and these tales aren't stories so much as invented conversations that these men could theoretically have had.

While several notable writers took on this story, including the philosopher Plato, it's Plutarch's that is the most famous.

Plutarch writes of Pittakos, one of the seven storytellers and lawmakers on the isle of Lesbos, recounting his own tale from his native island. Pittakos tells of seven Greek kings sailing to Lesbos to colonize it. Under the instruction of the oracle, they were to sacrifice one of their daughters. When the girl is chosen, a boy named Enhalos appears – he is in love with her and they jump into the sea, deciding to live there. He later reappears, flanked by octopuses, to save the islanders from a tsunami. Now a sea-dweller himself, his new natural habitat is represented even when he's on land by his octopus guard.

SCYLLA

We've already heard about the Kraken from Norse mythology, but Greek mythology has a near identical giant monster, the Scylla. The physical form of the Kraken differs in various cultures, including Norse mythology where it's sometimes a squid rather than an octopus. The Greek Scylla has even more variations in her depicted forms. She is sometimes a six-headed fish-cum-dog, sometimes she has one head with a fish tail and a few dogs sprouting from her torso and other times she more closely resembles a giant octopus.

Scylla is said to live on one side of a narrow stretch of water (generally assumed to be the Strait of Messina) with her fellow sea-monster counterpart, Charybdis, living on the other. Sailors that attempted to pass through the strait ran the risk of encountering either one of them should they sail too close to either side: Scylla could eat them while Charybdis could sink their ship with the whirlpools she created.

Scylla appears in Homer's *Odyssey*, Ovid's *Metamorphoses* and plenty of other stories and ancient myths.

OLD STORMALONG AND THE GIANT OCTOPUS

We'll stay in the lands of the fictional for a moment more with a modern tall tale. Tall tales, or tall stories, are usually greatly exaggerated stories that contain an element of truth. While this one is pure fiction, it originates as a legend told through sea shanties in Rhode Island.

Captain Stormalong is said to have washed up on a beach as a baby, already incredibly long at 5 m (18 ft) before growing to 9 m (30 ft) tall. He appears in sea shanties – songs sung by sailors – as far back as the 1830s. He was the captain of an impossibly tall clipper ship, and his nemesis? The Kraken, an enormous, monstrous octopus.

There are plenty of versions of Old Stormalong's life but in one, he tried to kill the Kraken and was so upset that he failed, that he went inland to work as a farmer. Other tales have him wrestling an octopus that had caught hold of his ship's anchor after it broke free at sea and landed on the sea floor.

MOVING ON FROM MYTHOLOGY

The octopus has well and truly staked its spot in mythologies from around the world. From the ancient Norse Kraken to the guilty Te Wheke o Muturangi fleeing angry Pacific Islanders, our eight-limbed friend has clearly sparked imagination for hundreds upon hundreds of years. While the myths are filled with wonder and larger-than-life characters, there's much interest and even disbelief to be found in the octopus stories that hold truth or are even entirely real.

Now we'll move this chapter closer toward the incredible truths and mysteries that surround octopuses, showing that you don't even need a myth to have a really great story concerning this savvy cephalopod.

THE ARGONAUTS FROM LEGEND TO REALITY

Let's now step into a legend. Legends are usually based on some truth and in this area of storytelling we find the argonauts. Now, funnily enough, argonauts are a real species of octopus and an unusual one. They're pelagic, which means they stay in the water rather than on the seabed. They're also very small, less than 10 cm (4 in.) in length for females although their shells can be much longer at 30 cm (12 in.). The males are much smaller, around 2 cm (⅘ in.).

What makes argonauts unique is that they have a "shell" of sorts, an incredibly thin sac that they tuck themselves inside of. The Greek philosopher Aristotle documented that these tiny creatures, often called paper nautiluses, sat in their paper-thin shells and used dorsal "fins" on two arms to catch the wind and sail along the sea's surface – not unlike the Argonauts from the Greek myth, *Jason and the Argonauts*. Aristotle's ideas became legend – a legend that lasted for two thousand years!

For a long time, scientists and naturalists had no idea if the argonauts created the shells themselves or adopted

them like a hermit crab. They also couldn't work out why the shells became larger when they reproduced. It wasn't until the nineteenth century that the secrets of the argonaut were discovered, and they were revealed by a fascinating person.

THE WOMAN BEHIND THE PAPER NAUTILUS

Jeanne Villepreux-Power was just 18 when she walked over a hundred miles through rural France to take up a seamstress's assistant position in Paris in the early 1800s. By 23, she was in love with and married to an English merchant and living in Sicily. Although women of the time had little to no access to scientific education, she set about learning natural history, particularly as it related to the island. She came across the paper nautilus, the argonaut octopus, and became fascinated. The uncertainty around why it had a sort of shell amazed her – no naturalists had thus far agreed on what it was for.

What Jeanne realized was that researchers had never studied living argonauts. They're hard to observe as they're extremely flighty, fleeing humans the moment they spot them. To observe them alive, she built a platform above a series of underwater cages. She

could then watch them from the water's surface. She abandoned her platform for being impractical and moved her research indoors, inventing the laboratory aquarium with the tanks she set up to house living argonauts.

Jeanne discovered that the female argonaut did indeed grow its own shell (the male argonauts were shell-free), and it happened soon after it hatched. Diligently observing hatchlings, she discovered that they had already tucked themselves up and were developing incredibly thin films around themselves. Over hours, she watched amazed as the films became thicker and furrowed. She also discovered that these little octopuses used their dorsal arms (Aristotle's sails) to repair any holes in their shells as well as using found shell fragments to patch up gaps – revealing stunning intelligence.

THE MYSTERY OF THE LIGHTS

In two separate places, in an aquarium in Coburg, Germany, and in a marine lab at the University of Otago, New Zealand, the lights started mysteriously breaking down. This is the story of what happened in Germany.

The Coburg aquarium's director recounted how staff turned up one morning to discover that the aquarium's electrical circuit had shorted. While aquariums have backups, it wasn't ideal, and they had it fixed. The next day they found the same thing had happened, no lights. The third day? The same.

Some intrepid staff decided to stay the night to find out what was causing the electrics to fail, and they caught resident octopus Otto red-handed (or red-armed). He had been climbing to the top of his tank and squirting water at the spotlight bearing down on him, shorting the circuit and dousing himself in darkness. Octopuses, it turned out, aren't fans of bright lights. Otto had been trained to squirt water at visitors for entertainment, but was using his skills in his own pursuits. Something similar had been happening in New Zealand.

PAUL THE PSYCHIC OCTOPUS

With octopuses being one of the world's most intelligent creatures and football being the world's most popular sport, perhaps it's inevitable that the pair should overlap at some point. No, the octopus in question never wandered onto a pitch. Instead, one particular eight-armed character called Paul was tasked with predicting the results of the 2008 Euro and the 2010 World Cup matches – with surprising accuracy.

Paul was a common octopus living in the Sea Life Centre in Oberhausen, Germany. His handlers gave him two boxes of food for each match Germany was playing, one box decked in the German flag and the other in the flag of the opposing team. The box that Paul opened and ate from first was deemed the team that he predicted would win the match.

This unusual job for an octopus wasn't what made headlines; what made the news was that he was correct 67 per cent of the time for the Euros and 100 per cent of the time for Germany's World Cup matches. Coincidence? Only Paul knows the truth.

THE NEW ZEALAND ESCAPE ARTISTS

Two separate octopuses living in aquariums in New Zealand made headlines after particularly crafty escapes. Inky, a common octopus, was described as an "unusually intelligent" octopus, even for his already smart species. He was living in the National Aquarium of New Zealand in Napier, North Island, when one day he disappeared overnight. In the morning, staff realized that his tank had been left ajar and theorized that he'd made an impressive escape down a 50 m (164 ft) drainpipe that led directly to Hawke's Bay.

In a more drawn-out escape, another octopus, Sid, was a savvy beast who occupied himself almost solely with escape attempts, this time from the Portobello Aquarium in Dunedin, South Island. Sid made several attempts, sometimes being caught in drains or with one foot (or rather, arm) out the door. Eventually, staff decided it was time to release him into the wild where he so clearly wanted to be.

OCTOPUSES FROM OUTER SPACE

It's not hard to conclude that octopuses are phenomenally special animals. Several species are exceptionally smart and creative problem solvers. Others have strange and incredible habits, like wielding weapons. Most die after reproduction and almost all have unbelievable shapeshifting and colour-changing abilities. There are few, if any, other animals on the planet that are so consistently jaw-dropping. So, is it possible that octopuses aren't from this planet at all?

In 2018, a group of scientists published a paper looking at the possibilities of octopuses coming from another planet. Their paper, "Cause of Cambrian Explosion – Terrestrial or Cosmic?", was published in the journal, *Progress in Biophysics and Molecular Biology*. It concerned the Cambrian explosion, a term that refers to a period of time around 540 million years ago (and itself lasting roughly 10 million years) when complex animals suddenly started appearing in the fossil record. Before that, few fossils of such life have been found. The Cambrian explosion saw vast diversification in life forms on the planet and increasingly complex animals evolving, or at least that's what the fossils appear to show.

There are many theories for why the Cambrian explosion began, but in this particular paper the speculation travelled further afield. To outer space. They put forward the supposition that the reason octopuses, who even millions of years ago closely resembled the smart creatures we observe today, are so unlike any other creatures is that octopus eggs arrived on the earth via comets that crashed onto the surface from space. Preserved in ice through space, the eggs hatched on Earth to fill our oceans with a new, alien species.

They also floated the possibility that rather than the eggs themselves, extraterrestrial viruses travelled here and infected similar, earlier animals, transforming them into the extraordinary octopuses we have today.

🐚 LIVING WATER PISTOLS 🐚

Octopuses are so clever that numerous aquariums have seen them redirect their jet propulsion abilities with a more playful purpose: squirting visitors in the face.

Octopuses propel themselves through the water in the wild by sucking liquid into the mantle cavity (the rounded part of an octopus, above the eyes) and then forcefully ejecting it using muscle contractions. This expels the water behind them and shoots their bodies forward. This allows them to move very quickly when swimming about or fleeing predators. In the aquarium, they can use it for other purposes.

There have been numerous reports from aquariums around the world of common and giant Pacific octopuses learning to use this jet propulsion to shoot considerable amounts of water out of their tanks.

ROCKY'S ANTICS

At the National Marine Aquarium in the UK, Rocky the giant Pacific octopus spent his time emptying considerable amounts of water onto employees as well as visitors. Whether he was doing it playfully or trying to send a message is something only Rocky would know.

OTAGO'S CHEEKY OCTOPUSES

At New Zealand's University of Otago, we've already seen that an octopus managed to turn out the lights by squirting water, but the university also had a staff member explicitly targeted. Every time she passed the captive octopus it would try to drench her; behaviour it didn't display around other staff members.

LUNA'S LESSONS

In 2017, Bristol Aquarium in the UK filmed their newly acquired giant Pacific octopus, Luna, squirting water at staff who were handling her. They took it to mean she was happy to see them rather than a sign of anti-social behaviour. It was clear by her inquisitive arms that she wasn't grumpy. This could show that octopuses use their jet propulsion abilities to display a range of feelings, depending on the situation.

🐚 A DISCERNING EATER 🐚

Perhaps one of the most wonderful stories that demonstrates the personalities of octopuses is that recounted by Jean Boal, an animal behaviour expert and marine biologist. When Jean was working in a lab housing a number of octopuses, she would walk alongside the tanks and drop thawed squid into them to feed each octopus. The lab would buy this food frozen in bulk as it was cheaper than freshly caught food.

One day, Jean was feeding the octopuses when she turned around to see one holding its bit of squid and watching her. She returned the gaze. It maintained eye contact with her while making its way through its tank and then, very deliberately, dropped the squid down the tank's outflow pipe, discarding it. Jean understood that the octopus was giving her a very clear message: this food is not up to scratch. Of all the incredible behaviour she's witnessed in her career, she reports, this is one of the moments that truly stood out as a marker of octopus intelligence. And presumably attitude.

THE OCTOPUS WHO
FLOODED HER AQUARIUM

In 2009, an octopus at Santa Monica Pier Aquarium decided to try out a new skill: mischief. The aquarium had recently had brand new floors installed, and one night, the captive octopus took apart one of the valves in her tank so the pipe flowed onto the floor rather than into her tank. This caused around 7,570 litres (200 gallons) of sea water to spill out and flood the room and adjoining offices. It was estimated that the water was flowing for around 10 hours before staff stopped it. The octopus was fine and watched the clean-up from her tank.

As aquarium staff are well aware, once an octopus learns how to do something, they'll do it again for fun, for escape, for mischief, for anything. The handlers at Santa Monica were careful to reinforce the valve the octopus had taken apart to prevent her from causing such chaos again. The aquarium's outreach coordinator said that while up until that point the octopus didn't have a name, after the flood, some people were suggesting they name her "Flo".

AN OCTOPUS'S
FACIAL RECOGNITION

If the story about an octopus squirting water at a specific aquarium employee is true then octopuses must, to some degree, be able to recognize individuals and remember them. But what does the science say?

In one study, eight giant Pacific octopuses were approached by two people, the testers, who they'd never met before. Over the course of two weeks, one of the people fed the octopuses every day and the other stroked them with a bristly brush, a feeling that would've been mildly annoying to the octopuses.

The testers made sure that the octopuses saw them each time and also recorded various factors such as the octopuses' breathing rates, skin colour, behaviour, etc. This would allow them to see if they began behaving differently as they adjusted to each tester's presence, indicating that they began to recognize them and react differently to them. By the second week, there was a clear difference, particularly with movement. The octopuses tended to move toward the tester who fed them and away from the tester who annoyed them with a brush.

This result may seem obvious, after all, we witness similar behaviour in dogs and cats as well, who quickly

recognize individual people and respond accordingly. But remember that dogs and cats are complex mammals and known to be very intelligent. Octopuses, while smart, are a completely different type of animal and much of their behaviour is still poorly understood. Studies such as this help researchers learn more about how octopuses view the world, how they respond to things around them and how we can better understand them in general.

Finally, going back to water squirting, the octopuses in the trial tended to squirt water at the brushers and not at the feeders, suggesting that they do indeed use water jets as a deterrent when in captivity.

CHAPTER FOUR:
OCTOPUSES IN CULTURE

We've learned what makes octopuses some of the most fascinating and almost unbelievable animals in the world so it's no surprise that they've infiltrated popular culture as much as they have. Octopuses crop up everywhere from children's books to iconic songs, representing everything from mystery and knowledge to multitasking and brain power. In this chapter, we'll take a look at some of the places you can find octopuses in books, films, music and even engineering. After this, you'll probably start noticing their influence wherever you go – it really is amazing where they turn up!

🐚 NAMING THE OCTOPUS 🐚

The octopus got its name from the Ancient Greeks and is Latin in origin. It's a compound word that's formed of *oktō* (eight) and *pous* (foot). While you may have heard people refer to the plural of octopuses as "octopi", it's actually incorrect as it uses the wrong form of the Latin noun. So, you can be sure that while some may say "octopi", you'll always be right when you say "octopuses".

Perhaps even more interesting is the name of the octopus's cousin, the squid. A squid's Latin name is *calamarius* and comes from the Ancient Greek and Latin words for "pen" and "pen case". This is presumably due to the thin shape of the squid and its habit of squirting ink. The common English name "squid" has unknown origins although it's believed to have appeared first in the sixteenth century. "Squid" was developed by sailors, possibly as a corruption of the word "squirt".

ARISTOTLE'S OPINION ON OCTOPUSES

We've already heard that the Ancient Greek philosopher Aristotle had some interesting ideas on the locomotion of the argonaut but he also had a fairly low opinion on the mighty octopus. During his marine observations off the Greek island of Lesbos, he made some exceptionally forward-thinking notes including correctly identifying that one of the male octopus's eight arms was used in reproduction. However, he also observed that "the octopus is a stupid creature for it will approach a man's hand if it be lowered in the water".

To Aristotle, an animal that came toward a person rather than stay away was unintelligent because it was endangering itself. Of course, what Aristotle mistook for stupidity was extraordinary intelligence and, above all, curiosity. Octopuses will flee from danger, but they also are intrinsically drawn to the novel, always wanting to investigate something new in their environment. It's this desire for knowledge that makes them so endearing, as well as vulnerable.

MINOAN OCTOPUS POTTERY

The Minoans were an ancient culture based on the Greek island of Crete and you might be a little familiar with them already. Ever heard the Greek myth of Theseus and the Minotaur? Theseus was Minoan. This culture spanned around a 3,000-year period between 3100 and 1100 BCE and while there's still much mystery around their society, their elaborate art – namely, their pottery – has survived remarkably well.

Minoan pottery developed significantly around 2000 BCE with the adoption of the potter's wheel. They created what became known as Kamares wares, which were pots adorned with beautiful artworks painted in light colours on black backgrounds. Around 500 years later, the marine style of pottery was established, and this is where our friend the octopus arrives. The style which features our friend tended to have a light background with sea creatures expertly painted in black. Such vases and pots decorated with octopuses were common in elite circles. You can see these ancient vases today online and in Cretan museums.

INK ON THE SKIN

Given that octopuses use their own ink, perhaps it's apt that they're also the subject of plenty of tattoos. Body art and the sea have gone together for hundreds of years and tattoos have been prevalent in many cultures around the world, including those of European and North American sailors from at least the sixteenth century.

Thanks to the myriad abilities, skills and legends surrounding octopuses, they make for fantastic tattoo subjects. Octopuses are complex animals with a serious slice of the enigmatic about them, something that many people closely relate to. Wisdom is a common meaning for an octopus tattoo, for reasons that should be clear by now!

Other meanings an octopus inked on the skin can embody include strength and resilience, mystery and depth, and transformation or rebirth.

Tattoos featuring these eight-armed creatures also have varying symbolism throughout the world, often linked to local legends and traditions. In Polynesian culture for instance, octopus tattoos may represent a spiritual connection, while in Japanese culture, they could represent power or be a symbol of protection.

PLINY THE ELDER'S ENORMOUS OCTOPUS

When it comes to ancient literature and animals, it's hard to avoid mentioning *Natural History* by Pliny the Elder. The most comprehensive work of "non-fiction" from the ancient world, Pliny wrote and compiled it during the first century ad. Encyclopaedic in form, gleaned from the expert knowledge of others, it contains entries spanning the animal and plant kingdom. Some entries are fascinating and true; others, not so much.

In it, Pliny recounts a story of a giant octopus that it is said repeatedly broke into a garum factory in Southern Spain during the night to eat fermenting fish. Garum was a fermented fish sauce, much like those used today in Asian cuisine.

The factory workers, it is told, built a huge fence around the buildings to protect their produce from the enormous octopus but it merely climbed over it. Eventually, the workers tracked the octopus down with dogs and killed it. Not a story you might find in today's encyclopaedias, but this story is a fascinating insight into the stories that were passed around in ancient times.

TOILERS OF THE SEA

Ever since pen was first put to paper, octopuses have been represented as monsters in literature, serving to make the clever creature more misunderstood. One of the best modern examples in relatively modern times is in the work of French novelist, poet and playwright, Victor Hugo.

Victor Hugo (of *Les Miserables* and *The Hunchback of Notre Dame* fame) published *Toilers of the Sea* in 1866, which directly proceeded *Les Miserables*. In it, his protagonist plans to rescue a wrecked ship in order to marry its owner's niece. During the rescue, he fights a monstrous octopus (sometimes translated into English as a squid) that's already claimed the ship's captain.

Hugo referred to the octopus as a "devil fish" that sucked the life out of its victims, cementing the idea of octopuses as monsters into readers' heads. This wasn't the first depiction of monstrous octopuses. Tales of the Kraken had already appeared in Norse mythology and in the English translation of the *Natural History of Norway*, in which a Danish author listed such mythological monsters as fact.

THE SOUL OF AN OCTOPUS

In 2015, a book which transformed how many of us saw octopuses was perhaps a little responsible for the increase in octopus-related media over the following decade. *The Soul of an Octopus* is a memoir by American author Sy Montgomery and was a *New York Times* bestseller and a finalist for the National Book Award in the USA.

Montgomery wrote the book after her *Orion* article, "Deep Intellect", about an aquarium octopus went viral. A natural historian and journalist, the bulk of her work throughout her career has taken her to far flung places and up close with animals from gorillas and pigs to the other worldly octopuses.

The memoir tells of Montgomery's search to understand more about octopuses and how they think and behave. She visits aquariums, talks to experts and interacts with octopuses in captivity and in the wild. An effusive book, Montgomery's love for octopuses is undeniable from the first page. It's impossible to read her book and not fall in love with these creatures.

THE OCTOPUS

This 1901 novel isn't about an octopus at all, nor the sea in any shape or form. It's this that makes the title so interesting. Norris's classic novel tells of a dispute between a railway company and wheat farmers in California. Based on a real tragedy in 1880 that occurred over disputed land between the two, the novel tells of the evils of monopolies.

By using an octopus as a reference in the title, Norris is referring to the railway's all-consuming and powerful grip on the farmers. An octopus has eight arms so is capable of controlling many things at once, a powerful metaphor for a monopolistic organization unfairly controlling the lives of independent people.

We've all heard metaphors relating to octopuses, often in good humour when someone is perhaps carrying too many things, but you can also find octopus metaphors in organizational design and teamwork theories, where the concept of central and distributed thought is useful. This novel is a great example of how this creature is so notable that its identity transcends its environment.

THE NOVEL THAT CAUGHT EVERYONE'S ATTENTION

Octopuses in stories are usually monsters or cuddly childhood characters but in 2022, Shelby van Pelt brought a new eight-armed figure onto the literature scene. Her octopus was a protagonist.

Remarkably Bright Creatures tells of a lost young man named Cameron and a widow, Tova, who works as a cleaner in an aquarium where Marcellus lives. Marcellus is a giant Pacific octopus, one of the most intelligent species, although he has little desire to interact with his keepers. But Tova and Cameron are different and the three strike up an unusual friendship. Part of the novel is told from Marcellus' point of view and it's this that makes the already wonderful novel so remarkable. With a full personality (and quite a grumpy one), Marcellus is the star of the story, helping Tova and Cameron heal from their troubles.

Becoming a bestseller, the novel went on to sell millions of copies, feature as a book group read around the world and spark a larger interest in and appreciation for octopuses.

OCTOPUSES IN NON-FICTION

After readers turned the last page of *Remarkably Bright Creatures*, where did they go to indulge their newfound love of octopuses? To some outstanding modern non-fiction that helps explain even more just how wonderful these animals are.

OTHER MINDS

Shortlisted for the 2017 Science Book Prize, *Other Minds: The Octopus and the Evolution of Intelligent Life*, offers a fascinating deep dive into how octopuses evolved and why they became so smart. It not only helps readers understand everything octopus, but it also sheds light on evolution in general and how close and far we are from these incredible creatures.

MANY THINGS UNDER A ROCK

David Scheel is a professor of marine biology at Alaska Pacific University and is undoubtedly an octopus expert, having studied them for over two decades. He distils his knowledge and passion into this book in an accessible way that's made it a bestseller, often describing scenarios from an octopus' point of view.

SECRETS OF THE OCTOPUS

After writing *The Soul of an Octopus*, Sy Montgomery went on to write *Secrets of the Octopus* as a companion to a National Geographic documentary. This book, published in 2024, offers fascinating insights into their world.

OBSESSIVE ABOUT OCTOPUSES

Being fascinated by octopuses isn't just for the adults, there are some non-fiction books about the marine world for children too. Covering the octopus and its ocean peers, the 2020 *Obsessive About Octopuses* gives children facts to blow their minds as well as showing why it's so important to take care of our seas and their inhabitants.

DOCUMENTING A STRANGE AND BEAUTIFUL FRIENDSHIP

If there was anything that brought the marvels of octopuses to the forefront, it may well have been the breakout 2020 Netflix documentary, *My Octopus Teacher*. On the face of it, this documentary looked quite simple and pretty niche with no one realizing how big it would become.

The story follows filmmaker Craig Foster as he gradually strikes up an unlikely subsea friendship with a common octopus. Free diving in the cold kelp forests off False Bay near Cape Town in an attempt to recover from burnout, Foster was no stranger to the area. In 2012, he founded a non-profit organization, the Sea Change Project, to protect the kelp forests, an incredibly important part of the marine environment.

One day while diving, he discovered a common octopus and found it so fascinating that he decided to return to her den every day to learn more about her behaviour. If Craig had simply made a documentary on octopus habits then it might not have quite had the impact it did, but he didn't. Instead, in the piece the octopus begins to trust Foster, getting used to his presence and reaching arms out to him, investigating this new face in her environment.

The documentary follows her life and their interactions. It shows her defending herself (sometimes not very successfully) against her main predators, pyjama sharks, as well as learning rapidly and becoming at home with her human companion.

The documentary follows the octopus' entire life and Foster's emotional journey with her as well, deepening his connection with the ocean as well as his son, who he takes to see the octopus.

My Octopus Teacher won streams of awards and was nominated for even more, becoming a surprise hit around the world. It also helped change people's perception of these mysterious, alien-like animals and helped transform their reputation from ocean monsters of myth to intelligent, innovative animals that deserve our protection.

THE OCTOPUS IN MY HOUSE

In 2019, the BBC released an episode of *Natural World* called *The Octopus in My House*, which follows author and octopus expert David Scheel as he brings an octopus to live with him to study its behaviour. Author of 2023 *Many Things Under a Rock*, Scheel introduces a day octopus, Heidi, into a new tank in his home and begins his observations.

Soon, Scheel and his daughter form a strong bond with Heidi and spend considerable time playing with her, giving her toys and watching how she invents her own games. Heidi becomes a part of their family, arriving in the wake of Scheel's divorce, and refocuses him on the new project of studying and caring for his new housemate. The documentary offers an incredibly in-depth view into the charm and problem-solving abilities of an octopus. The episode also goes into considerable detail about octopus behaviour in general both in the wild and in captivity, and shows how even Scheel (an octopus expert) learned new things about how these creatures live every day.

THE OCTOPUS IN HORROR MOVIES

Documentaries like *My Octopus Teacher* and *The Octopus in My House* are important because they turn the "octopus monster" character on its head. And it's been used *a lot*. Countless movies use octopuses – usually impossibly giant ones – as their terrifying monsters. Here are a few examples:

Mega Shark vs Giant Octopus – A 2009 monster disaster movie about a prehistoric giant shark and octopus breaking free from ice and causing global havoc.

It Came from Beneath the Sea – A 1955 monster horror featuring a radioactive giant octopus attacking sites around San Francisco.

Tentacles – A 1977 creature movie about another giant octopus causing mayhem along an American coastline, released in the wake of the success of *Jaws*.

Monster – A 2008 Japanese Kaiju movie (a genre specifically involving gigantic monsters) that follows a giant octopus released from the earth after an earthquake. This movie was created to capitalize on the success of the monster disaster epic, *Cloverfield*.

THAT JAMES BOND MOVIE

If you think about a movie starring an octopus then perhaps the first one that comes to mind is the James Bond classic, *Octopussy*. It would be a little strong to say that octopuses themselves play a central role but Octopussy is a blue-ringed octopus, tiny but extremely venomous and a real species.

This little octopus was first owned in the story by a jewel thief, Dexter Smythe. Later it is taken on by his daughter, who he also named Octopussy, who becomes a central character in the story. Octopussy, the daughter, runs the Octopus Cult and inherits the tiny octopus – an unlikely situation as octopuses only live for around a year or two, but hardly the most unlikely situation in any given Bond movie!

In classic Bond style, there are car chases, bombs on the brink of explosion and stolen jewels. The octopus was much larger in the film than a real life blue-ringed octopus and was portrayed as more aggressive.

TWENTY THOUSAND LEAGUES UNDER THE SEA

First serialized in a newspaper in 1869, the French writer Jules Verne's incredible science fiction novel was a success right from the start. A true classic of both science fiction and general fiction, the story follows a marine biologist on a deep-sea mission to track down a mysterious sea monster. When the sea monster turns out to be Captain Nemo's futuristic submarine, the protagonists are taken on a wild adventure through the seas. The story has been adapted both closely and loosely many times in audio, stage, film and TV, each time changing to some extent.

Cephalopods appear when the Nautilus, Captain Nemo's submarine, is attacked by "devil fish". In the book they're giant squid, but in many adaptations they've been gigantic octopuses. The 1916 movie adaptation is an amazing example of not just octopuses on film but of underwater filmmaking in its infancy.

John Ernest Williamson invented the photosphere method of underwater photography and then built an animatronic octopus for the film *Twenty Thousand Leagues Under the Sea*, creating a terrifying and innovative scene.

DOCTOR OCTOPUS

Spider-Man might be able to create webs but his first major villain had the power of an octopus: Doctor Octopus, or Dr Otto Octavius. An early character, Doctor Octopus (Doc Ock for short) was introduced in the 1963 *The Amazing Spider-Man #3*. A classic mad scientist character, Doc Ock fell victim to an accident in his laboratory and ended up with four independently moveable tentacle-like metal arms fused to his spine. He'd developed the arms to handle dangerous chemicals safely, but his character backstory tells of his life falling into disarray and leading to the accident.

An angry villain, he turned to crime and eventually formed the Sinister Six, a group of supervillains determined to take down Spider-Man. His creator Stan Lee said that while he usually created the character before giving them a name, for Doc Ock, the concept of the octopus came first.

THE CEPHALOPOD
APPRECIATION SOCIETY

If you wanted to know just how many people love octopuses and other cephalopods so much that they'll seek out a community, look no further than the Cephalopod Appreciation Society on Facebook. With over 140,000 members, it's an active group sharing everything from octopus-related artwork to videos of cephalopods in the wild and captivity.

Groups like this are vital to the continued thriving of octopuses because, while none of their species are endangered, the ocean ecosystem in general is at risk from human activity. Community groups that share their appreciation for any sea animal only help bolster the wider appreciation and care society has toward marine life, helping to protect it.

OCTOPUSES IN CHILDREN'S SHOWS AND MOVIES

This far into the book it's probably safe to say we're all huge octopus fans by now, but as much as these charming animals capture adults' attention, imagine what they do for children's. Octopuses often feature in children's media and just like in adults', they're sometimes positively portrayed and sometimes negatively, taking on the roles of villains.

OCTONAUTS

Octonauts is a BBC-created animated children's series about a group of eight animal adventurers who live in an underwater base and go off on adventures, studying the ocean. One of their number is Professor Inkling, a Dumbo octopus.

SPONGEBOB SQUAREPANTS

This hit TV and movie franchise features Squidward, a grumpy neighbour of SpongeBob's and an octopus, despite his name. His address is 122 Conch Street, Bikini Bottom and he's a giant Pacific octopus.

OSWALD

The Nickelodeon show *Oswald* was launched in 2001 and featured a sweet, caring, blue-coloured octopus as its central character. Living in an apartment, it follows his life as he chats to his friends and neighbours along with his hot-dog shaped dog.

DEEP

After a global disaster floods the earth, an old octopus named Kraken (after the monster in Norse mythology) saves a diverse number of sea creatures by creating a community in a deep part of the ocean. The movie centres on his grandson Deep, also an octopus, who gets up to a little mischief and must head off on a big adventure to set things right.

PENGUINS OF MADAGASCAR

This cute spy movie is part of the *Madagascar* franchise and involves a troupe of penguins kidnapped by a giant Pacific octopus named Dave. Dave's annoyed because he used to be the centre of attention at any aquarium he lived in, but has since been upstaged by the penguins.

FINDING NEMO (AND DORY)

One of the sweetest octopuses in this list, Pearl is a minor character in the Disney Pixar classic, *Finding Nemo*. A tiny pink flapjack octopus, she goes to school with Nemo before he's taken. In the spin-off, *Finding Dory*, there's Hank the octopus, who's reminiscent of a giant Pacific octopus and plays a larger role than Pearl did in *Finding Nemo*.

SHARK TALE

Featuring an all-star cast including Will Smith, Renee Zellweger, Angelina Jolie and Robert De Niro, *Shark Tale* is about a little fish with a big lie, as well as a vegetarian shark. Its resident octopus is a villain, Luca, and while he's not the smartest, he loves to play piano.

THE LITTLE MERMAID

A classic Disney movie loosely based on a Hans Christian Anderson story, *The Little Mermaid* features the sea witch Ursula, who's portrayed as a human/octopus hybrid. With only six arms due to production difficulties in the late 1980s, it's hard to say if she's an octopus or not.

THE PACIFIC NORTHWEST TREE OCTOPUS

There's one species of octopus that we haven't touched upon at all and that's the land-dwelling Pacific Northwest tree octopus. This intrepid and large tree-climbing octopus is, of course, not real but it has been used when testing the critical thinking abilities of children. It was originally developed as an internet hoax in 1998 and went on to be used in two studies, one in the USA and one in the Netherlands to test how well children between 11 and 13 judged the reliability of a website.

The results showed that the children were easily convinced by the hoax when it appeared on a spoof website that framed the animal as endangered and in need of conservation support. This was despite many children in both studies having already received media literacy training.

Humorously for adults in the know, the octopus was said to live in the Olympic National Forest and be on guard against its main predator, Sasquatch.

"OCTOPUS'S GARDEN"

During a particularly challenging time for Ringo Starr when he was playing in The Beatles, he stayed on a yacht in Sardinia with the British comedian Peter Sellers, an experience that would prompt him to write his second song and vocalize his last for the band.

When served calamari and chips one day, the captain of the boat described how octopuses build gardens with interesting rocks and shells they find on the seafloor. Perhaps he was referencing how they collect them to build out their dens and protect themselves from predators.

The idea captured Starr, who said later that the difficulties in the band made him want to live under the sea instead. It turned into the *Abbey Road* song "Octopus's Garden", released in 1969 and much loved by fans of the band. It's also an interesting example of an octopus reference that frames the animals positively, rather than as monsters.

THE NATIONAL RECONNAISSANCE OFFICE

The USA's National Reconnaissance Office (NRO) also took the power and intelligence of the octopus as inspiration although in this case it sparked some scepticism and a little controversy. When they launched a reconnaissance satellite, US-247, for radar imaging, it hit the headlines not because of its task or launch, but because of its logo.

The NRO had chosen an aggressive-looking gigantic octopus to represent the satellite, depicted as sprawling across the earth with the phrase, "Nothing is beyond our reach" accompanying it.

The menacing logo was seen as controversial as its launch followed the surveillance disclosures in 2013, which were leaked by whistleblowers to international media outlets and on to the public. These leaks showed far-reaching intelligence and surveillance operations by countries such as the USA, so the use of an all-powerful octopus and an apparently too-true motto was considered by many to be advertising potential overreaching of power.

OCTOPUSES IN ENGINEERING

We live in a world with lightning-fast internet, supercomputers and ever-improving AI and we already have so many robotic things, from self-service supermarket checkouts to robot vacuum cleaners bumping our feet while we're in a work video meeting.

As humans, we might consider that thinking is harder than moving. We might walk into the kitchen to make a drink while procrastinating about a particularly difficult email, and the act of walking is so easy it barely even registers that we're doing it. But in robotics, the opposite is true. With phenomenal computing power, the *thinking*, programming side of robots, isn't necessarily the biggest challenge; it's the moving side.

Getting a robot to walk smoothly and safely from one end of the room to the other is an incredibly difficult task, even more so if there are steps or obstacles in the way. What's easy for us is near impossible for them.

When a robot needs to move in the world or pick something up, the challenge often lies in the field of soft robotics. This is the area that deals with more flexible materials that can adapt to their environments, for instance, when picking something up that doesn't fit perfectly into a rigid "hand". When soft materials are used, these robots can also be useful in other environments, like dangerous or disaster-struck places (such as earthquake-affected buildings).

Just like humans have been used as inspiration for robots, so have other animals and the octopus is a particularly interesting one.

IT'S ALL IN THE ARMS

One study from Temple University in Philadelphia, USA, designed a small worm-like robot that used suction cups inspired by octopuses as its way of scaling vertical surfaces. Soft robots that could achieve this reliably in the future could help with maintenance inspections of high and hard-to-reach places, as well as surveillance.

100 PER CENT SOFT

Researchers at Harvard's Wyss Institute (which specializes in bioinspired technology) designed the first 100 per cent soft robot, containing no electronics at all. Dubbed the "Octobot", the octopus-inspired robot can move autonomously using chemicals as fuel. This means it has no cables, no hard parts and can move without external interaction.

TACTICAL TAPERING

Still at the Wyss Institute, in 2020 researchers published findings from their study looking into the use of tapered, suctioned arms as opposed to cylindrical arms. The team developed a soft robot that had the same tapered, suction-cup covered arms as an octopus, and tested how well it would grip, place and move objects. They discovered that the tapering of the arm had a significant positive impact on how well the robot arm could grip and move objects: it allowed it to bend in more flexible ways. The authors noted that not only does their research help further soft robotics, but it also helps further understanding of octopus biology.

It's inevitable that octopuses will keep inspiring researchers in soft robotics and other technologies like camouflage, sharing their phenomenal biology with the robots of the future. So far, when robots have run into problems, octopuses seem to have some answers.

BUILDING THE
OCTOPUS'S REPUTATION

As we've seen in this chapter, popular culture is octopus-obsessed. From the gigantic, tentacled monsters of disaster movies and wild stories, to the charming octopus characters in children's shows and in real life documentaries, these animals have had our attention for over a thousand years.

What's important is that whatever octopus-related media we consume, we remember that these are actual animals living in the oceans, seas and aquariums throughout the world. No octopus species is thought to be endangered but our actions and cultures can have a negative impact on them and their environments. Sewage spills, microplastics and other pollution that finds its way into the sea can all impact octopuses, damaging their physical surroundings and their food availability.

Hopefully with so much positive media emphasis on octopuses in recent years, their reputation is now such that we can work harder to look after the oceans and reduce humanity's impact on them and their ecosystems.

CONCLUSION

Many of us have enjoyed the soulful presence of a dog, the comical and intrepid behaviour of a cat and even perhaps the peaceful calm of a horse. Very few people have spent any real time with an octopus, but those that do report extraordinary feelings of connection and reciprocal curiosity. We hope this book has made you feel like you, too, have spent time in the company of a wise octopus.

You may now find that you spot octopuses and their attributes wherever you look, from adverts to logos, in jeweller's windows and in media headlines. When you spot one, take a moment to remember what makes them so wonderful and carry that thought with you into your day. They're endlessly curious, forging connections and inspecting things; they're fast learners, knowing that life is short and knowledge is useful; they're adaptable, temporarily adjusting to fit the task; and they're determined, they'll always persevere until they can finally open the jar!

THE LITTLE BOOK OF TREES
Felicity Hart

ISBN: 978-1-83799-633-9

Connect with nature's sacred roots with this enchanting exploration of the magick and mysticism of trees

Their stories are written in folklore, they are revered across cultures, and they stand as powerful sources of ancient magick; trees have always been a source of fascination to humans. Uncover the secrets of the forest with this bewitching guide to tree legends, symbols and spells.

Have you enjoyed this book?
If so, find us on Facebook at **Summersdale
Publishers**, on Twitter/X at **@Summersdale** and
on Instagram and TikTok at **@summersdalebooks**
and get in touch. We'd love to hear from you!

www.summersdale.com

IMAGE CREDITS